SHARK
COLORING BOOK

Get FREE printable coloring pages and discounted book prices sent straight to your e-mail inbox every week!

Sign up at:

www.adultcoloringworld.net

ISBN-13: 978-1530597802
ISBN-10: 1530597803

PREVIEWS:

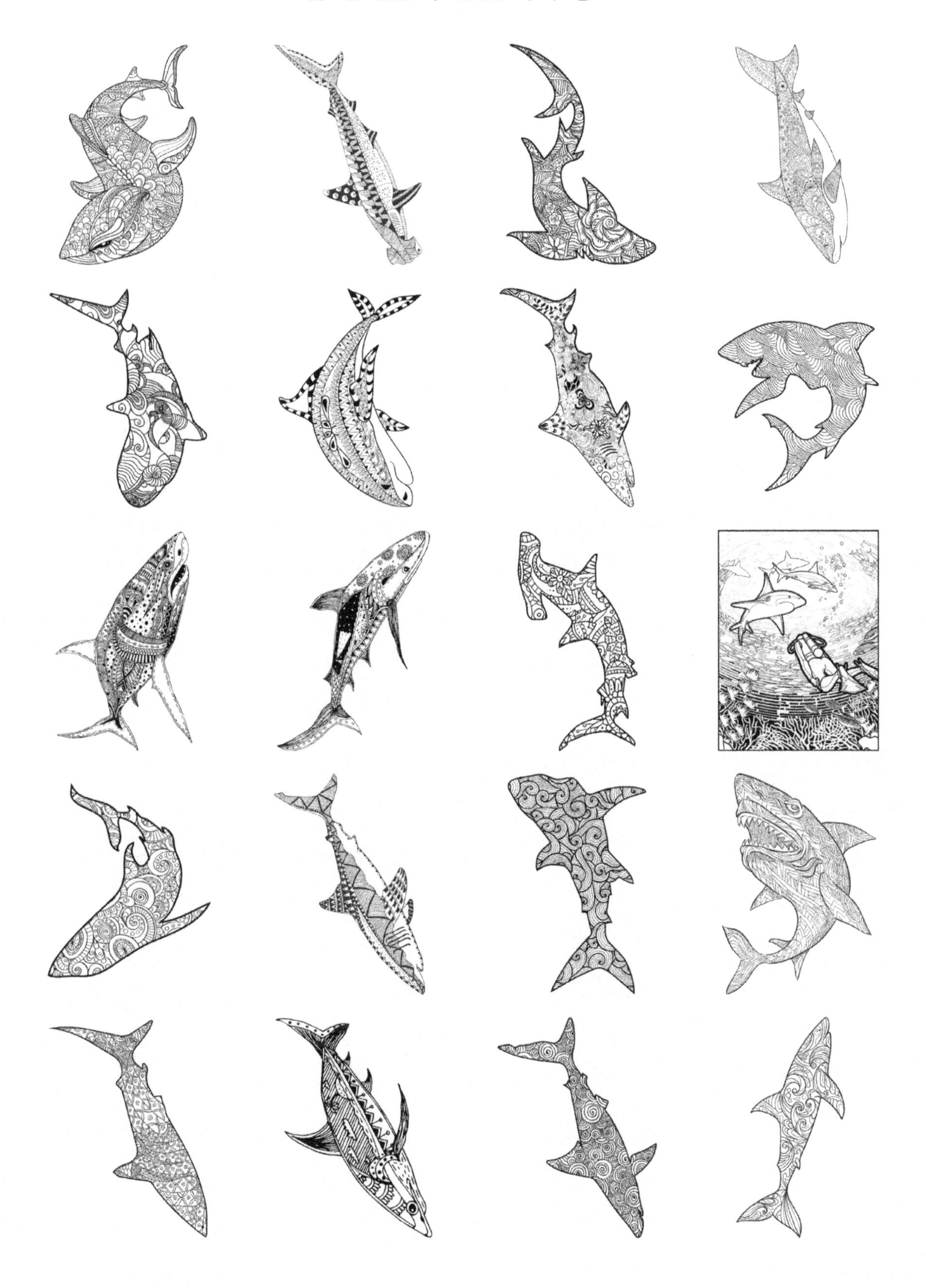

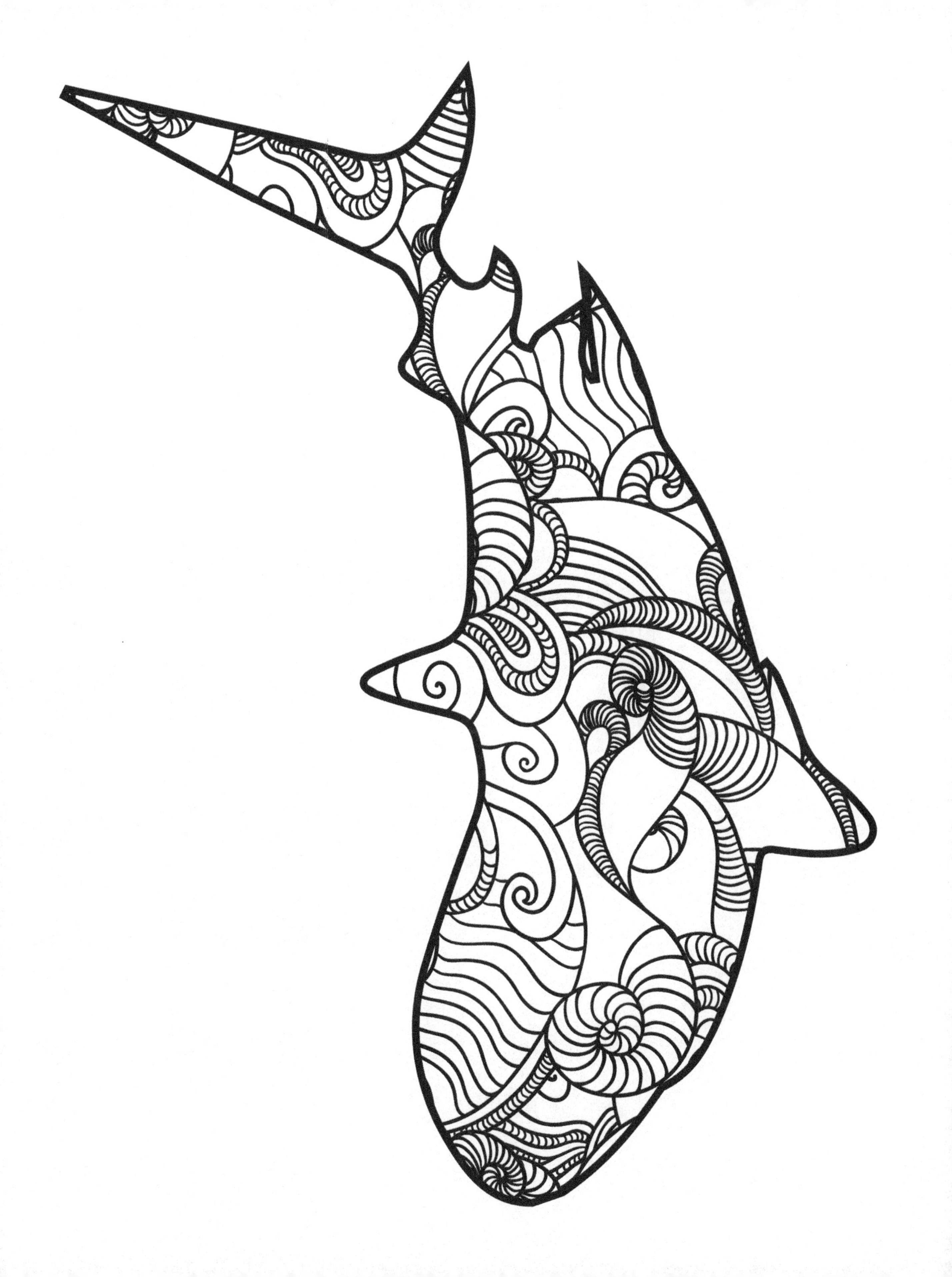

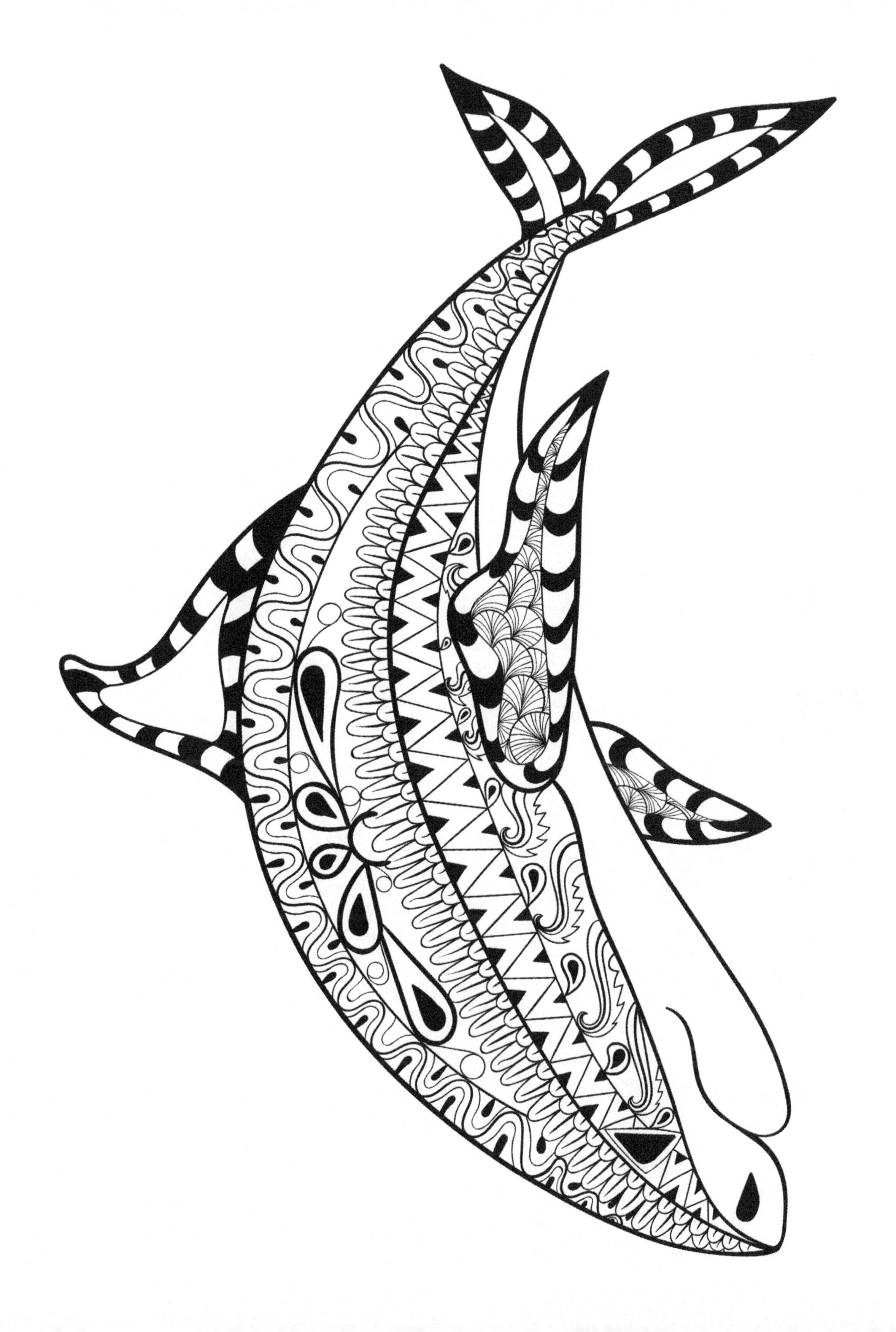

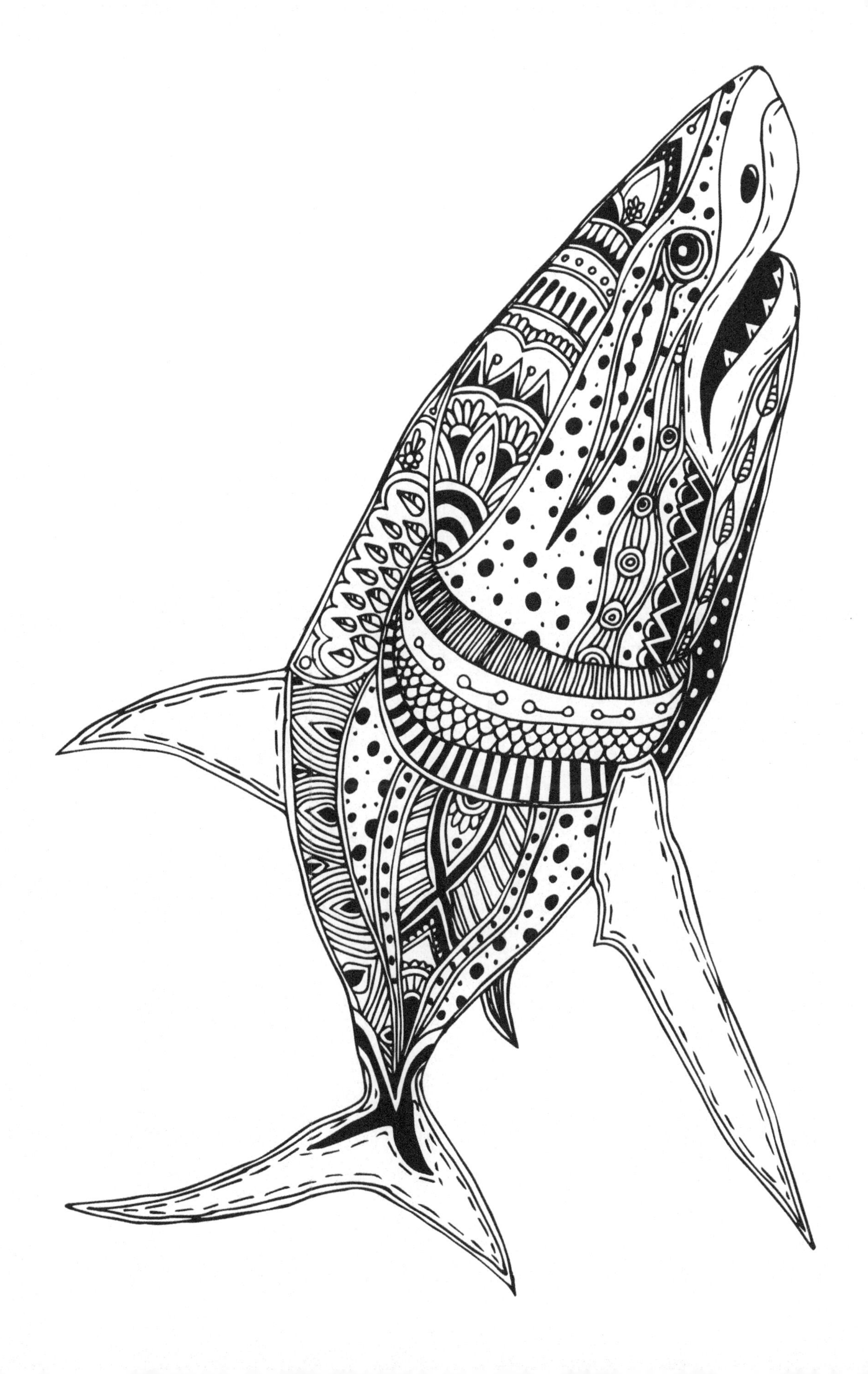

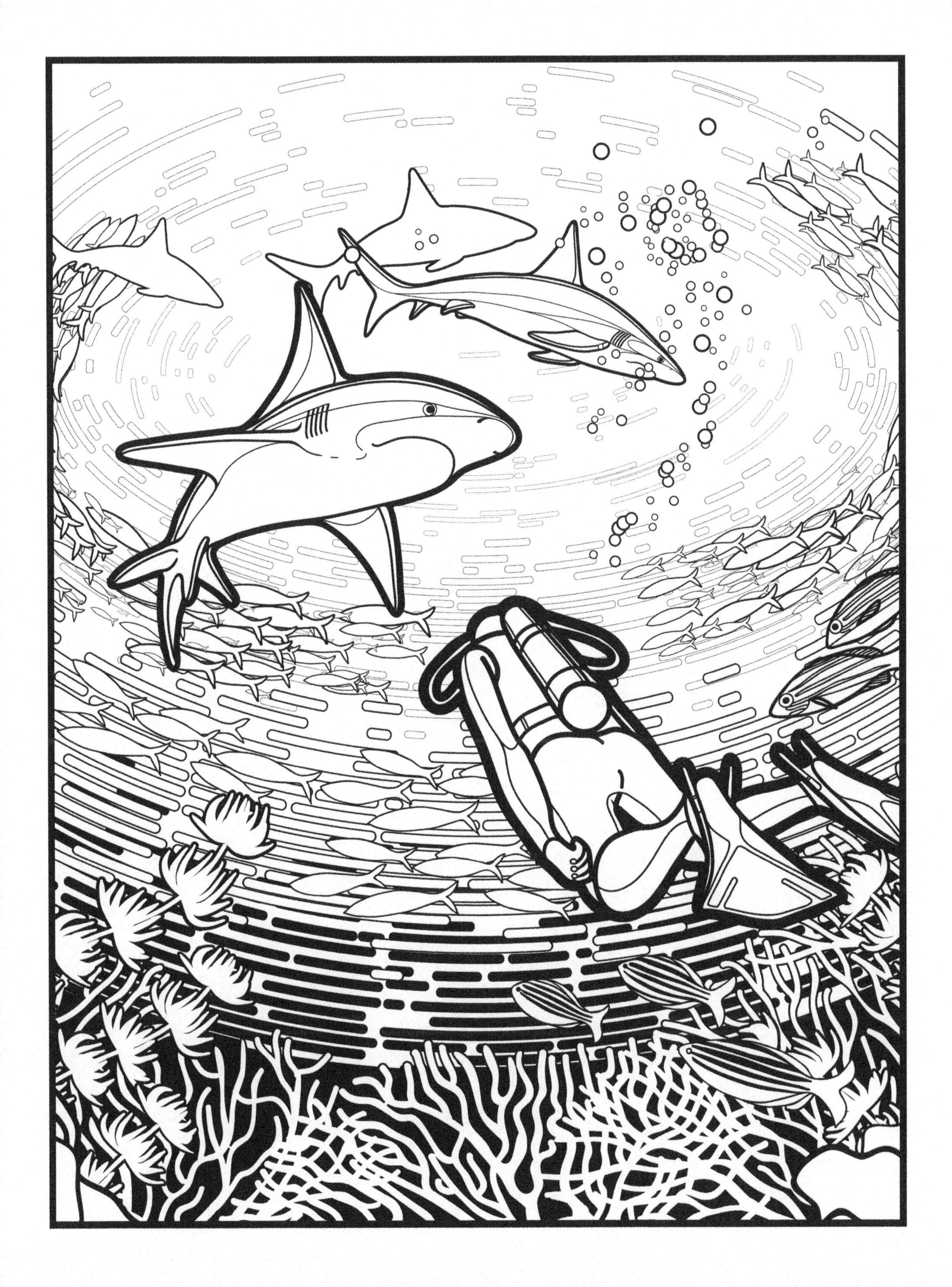

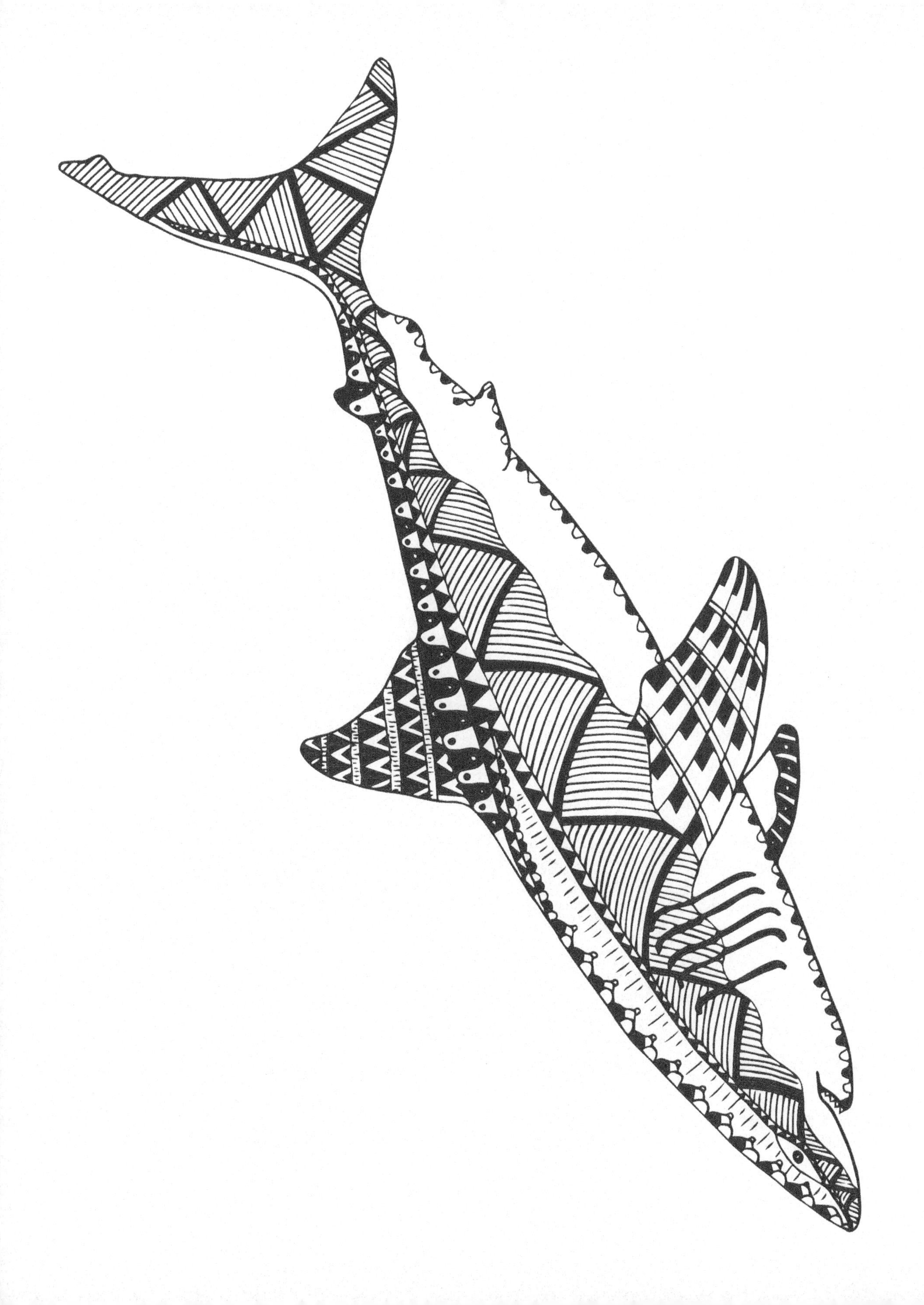

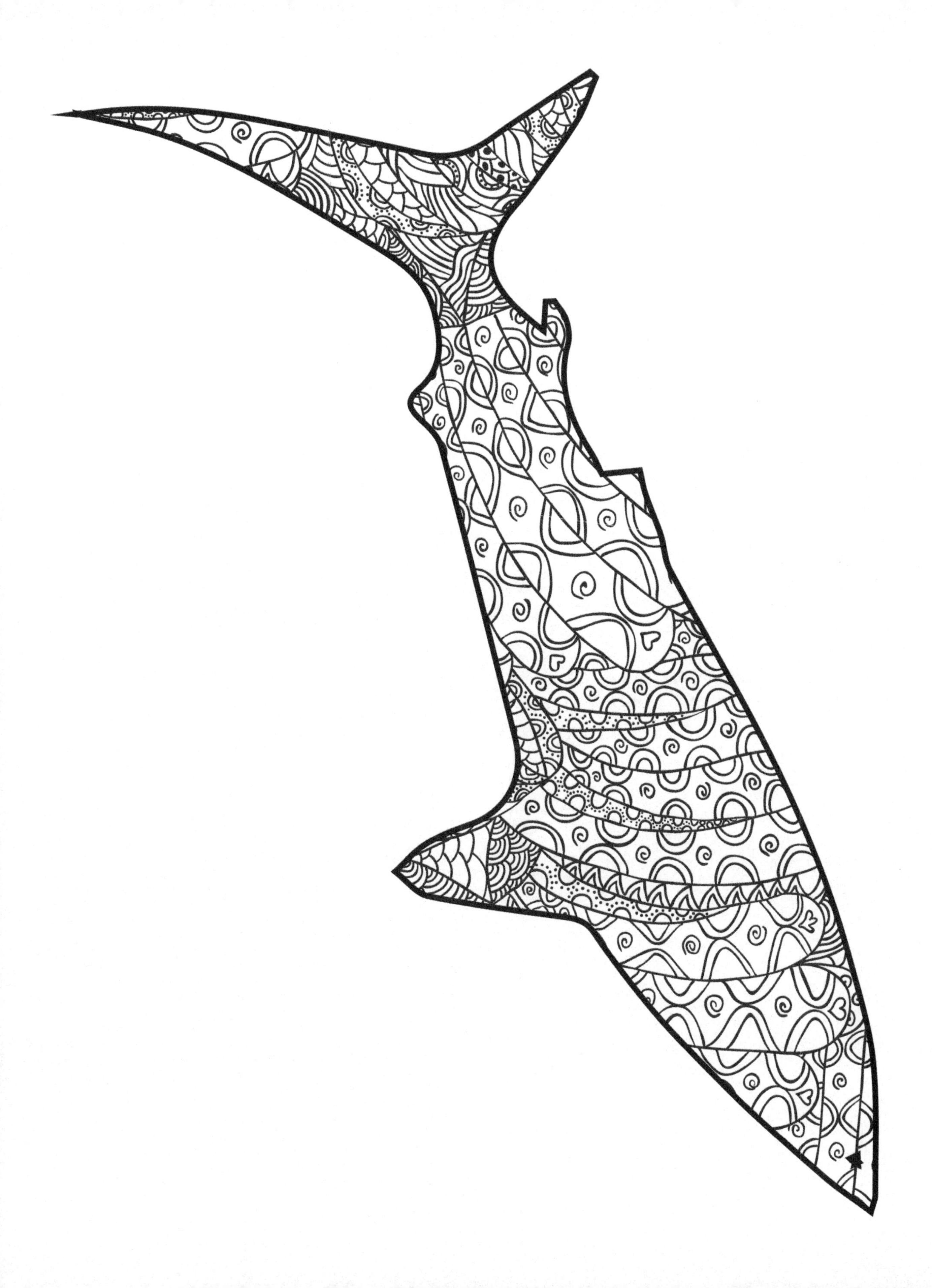

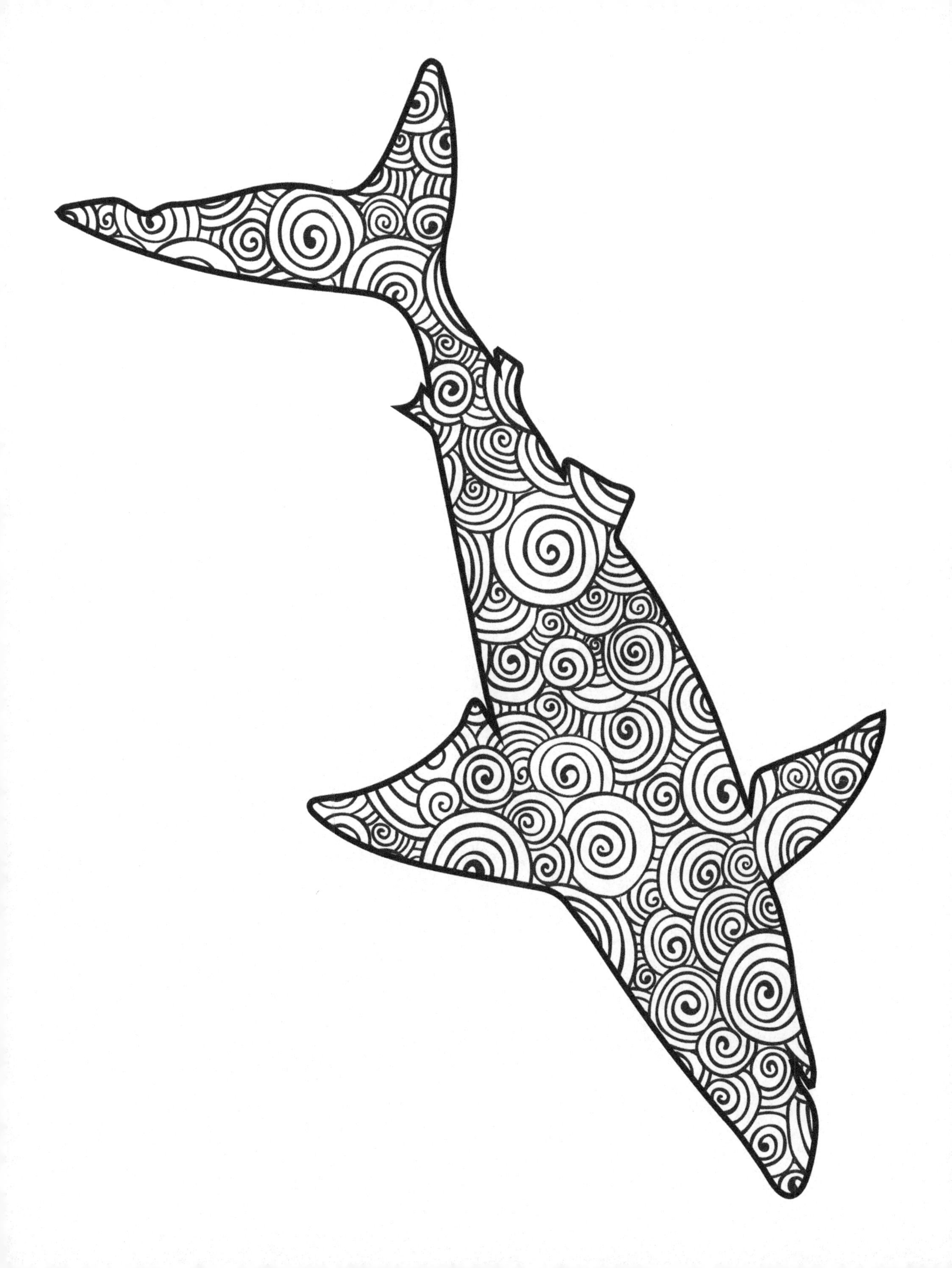

Color Test Page

Color Test Page

Made in the USA
Coppell, TX
14 December 2020